YOUR KNOWLEDGE HAS VALUE

Jan Seidel

Strategic Management in the Renewable Energies Sector

Using the example of a company which produces photovoltaic panels

GRIN Verlag

Bibliografische Information der Deutschen Nationalbibliothek:

Die Deutsche Bibliothek verzeichnet diese Publikation in der Deutschen National-
bibliografie; detaillierte bibliografische Daten sind im Internet über http://dnb.d-
nb.de/ abrufbar.

Imprint:

Copyright © 2007 GRIN Verlag GmbH
Druck und Bindung: Books on Demand GmbH, Norderstedt Germany
ISBN: 978-3-656-09261-2

This book at GRIN:

http://www.grin.com/en/e-book/77057/strategic-management-in-the-renewable-
energies-sector

BUSINESS MANAGEMENT
FOR THE BUILT ENVIRONMENT PROFESSIONALS

JAN SEBASTIAN FRIEDEL SEIDEL

Essay:

"STRATEGIC MANAGEMENT IN THE
RENEWABLE ENERGIES SECTOR -
USING THE EXAMPLE OF A COMPANY
WHICH PRODUCES PHOTOVOLTAIC PANELS"

School of the Built Environment, Heriot-Watt University

2007

Table of Content

List of tables and illustrations

1 Introduction

1.1 Strategic Management

There are various definition of strategic management in the literature because of the complexity and different approaches to deal with strategies. To give a brief idea what strategic management is all about the following definition is given. *"Strategic management is a process, directed by top management, to determine the fundamental aims or goals of the organisation, and ensure a range of decisions which will allow for the achievement of those aims or goals in the long-term, whilst providing for adaptive responses in the short-term"*. (COLE, 1997) Companies in the photovoltaic (PV) sector need to pass through the process of strategic management as well as their competitors. They will reach their aims and forecast or react on changes in the market to ensure the survival or development of the firm. This process will be presented in this work.

1.2 Photovoltaic and the Renewable Energies Sector

The need for strategies in this particular sector is getting demonstrative by the following statement:

"The enhanced importance of renewable energies is seen as one inevitable consequence of the acceptance Royal Commission on Environmental Pollution's (RCEP's) recommendation that the UK should attempted to reduce carbon dioxide (CO_2) emissions by about 60% from current level by 2050. An estimated 30-40% of electricity demand will need to be derived from renewable resources in order to achieve this objective, a target which implies a massive increase in development in renewable energy schemes".
(KELLETT, 2003)

The mentioned authors in this work have different ideas about aims and goals of this sector. Also the environment of PV producing companies is more or less discussed in academic papers and books in this field. The used schools of thought and various strategic analyses will be handled as well as the competitive advantages and future visions of companies in this sector.

2 The Process of Strategic Management

2.1 Purpose, Goals and Objectives

Companies set out their purpose, goals and objectives to achieve them by creating and implementing strategies. According to COLE (1997), the purpose or mission statement should define the key arguments of the company's business and behaviour towards their external and internal environment. "[They]... *are intended to provide a vision of why the organisation exists, where it intends to operate and how it intends to achieve its goals*" (COLE 1997). For a company which produces PV panels these statements could include the ideas of SHEFFIELD (1999). He reports that the demand of the increasing energy use must be meet by replacing fossil energies against renewable energies to save fossil resources and to minimise greenhouse-gas emissions.

In accordance with COLE (1997) the aims and goals are more focused on the success-criteria of the company's operations. A realistic goal could be correspondent to VAN OVERSTRAETEN ET AL (1996), "*[...] to obtain a power generation system which is able to convert solar energy into electricity at a cost/kWh which is comparable with other generation or delivery schemes [...]*". The objectives describe the intentions of specific units within the company. SHEFFIELD (1999) thinks that the companies need to improve the efficiency of the energy production of the PV cells and arrays and lower the cost by developments to be competitive in the sector of energy production. The following hierarchy classifies these goal-setting activities under the terms of COLE (1997):

Fig. 1: Hierarchy of goal-setting activities

2.2 Analysing the External and Internal Environment

After thinking about purpose, aims and objectives the PV producing company has to give thought to their external and internal environment before dealing with relevant strategies. According to WELFORD and GOULDSON (1993); external and internal stakeholders influence the process of planning and changing the future of the company. Critical key issues of internal influences should be identified like organisational, cultural or financial factors. The mayor differences to companies from other sectors in the Construction industry become explicit by having a look at the external influences. WELFORD and GOULDSON (1993) say that the government is a main factor because they *"[...] are increasingly applying market instruments to achieve environmental objectives"*. In Germany for example the government supports the construction of building integrated PV systems which are connected to the local energy grid. Agreeable to WEISS ET AL (2006), they engage the power companies to pay the costumer for the energy which they supply back to the grid. Another mayor external factor is the technology. PATEL (1999) explains that the technology of PV cells and modules is still in a developing process and new improvements come up regularly. The competitors in the renewable energy sector are standing out as well, but this will be discussed in an eventual chapter. Other important external influences are shown in the following figure which is correspondent to COLE (1997):

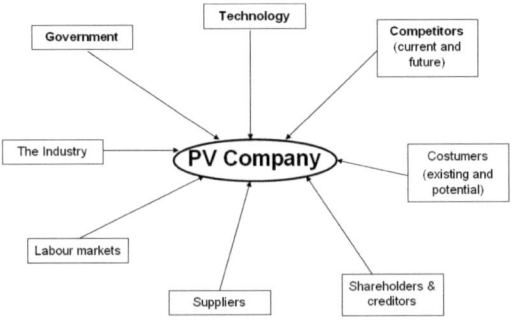

Fig. 2: External influences on a PV producing company

2.3 Schools of Thoughts

This chapter should explain which schools of thoughts could be relevant for a PV producing company to define their strategies. It will be discussed later which of them are temporarily in use by strategists of these companies. According to CLEGG ET AL (2005), the Chandler School says that the strategy should react on changing in the environment that affects the structure of the company. See that the potentialities of integrating PV in everyday life are still not exhausted; this could be a way of focussing on a strategy to enter new markets, in keeping with FVS (2003). The Strategic Planning School argues that strategy would be rational planning which means that decisions are made by *"[...] extrapolating from the past to plan the future"*. (CLEGG ET AL, 2005) Analysing the company in terms of strengths, weaknesses, opportunities and threats is ancillary to the Design School. Referring to that, strategies can be defined by audit the external and internal environment. CLEGG ET AL (2005) reports that the Positioning School is about to station the company in a competitive environment. As discussed before, the competitive advantage is quite big in the energy sector. The Five Forces Model of PORTER (1980) visualises that the rivalry among the competitors depends on different factors.

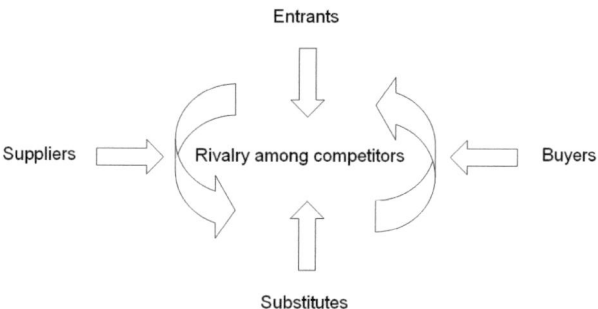

Fig. 3: Porter's Five Forces Model

If PV producing companies play around with future scenarios then this falls under the School of thought called Scenario Planning. Forecasting the future about, for example, what types of PV cells will be the most accepted or which kind of integration possibilities will be the most effective is a topic in an separate chapter. The mentioned Schools of Thoughts are just a selection of various opportunities and often combination should be used to consider all possibilities. But they are basic to think about strategies. They describe the different foundations to build up the processes of strategy.

2.4 Strategic Analysis

Strategic analysis takes up a central role of the strategy process. It is a dynamic way of handling information about the internal and external environment of the company to cause and react on changes. An example of this recurring process is shown in the following figure under the terms of LAURENS (2004):

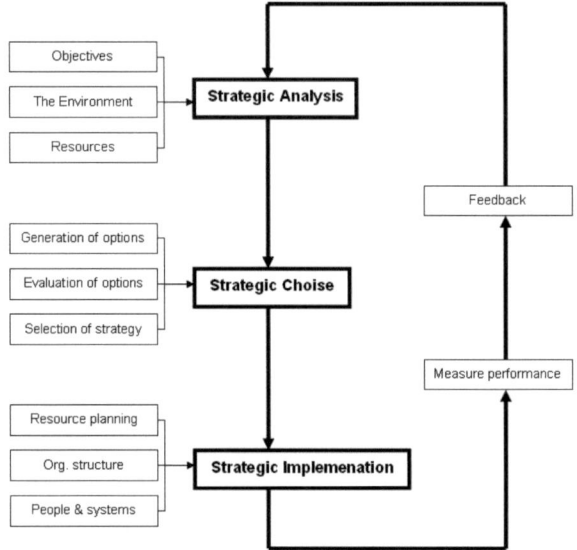

Fig. 4: A view of the strategy process

COLE (1997) writes about the BPEST (Business, Political, Economic, Social and Technological) analysis which enables strategists to *"[...] bring together a comprehensive review of the external environment [...]"*. As a result this analysis would spotlight positive and negative consequences of the company's future. Another technique which could be used by PV producing companies is the SWOT (Strengths, Weaknesses, Opportunities and Threats) analysis, discussed by COLE (1997). It should identify problems before they occur and demonstrate strengths of the company which could be capable of development.

A competitive analysis model is presented by MCNAMEE ET AL (1999). They say that companies should provide *"[...] individual firm reports so that participating firms can benchmark their performance in terms of measures such as: growth rates, internal performance measures, external performance measures and strategic priorities"*.

2.5 Competitive Strategies

Many authors who write about corporate strategies in the PV and renewable energy sector focus on the competitive advantage in the relevant markets. In accordance to VAN OVERSTRAETEN ET AL (1996) PV producing companies compete first of all with similar companies in the PV sector. Additionally, they compete with other technologies in the renewable energies sector like biomass, hydro power or wind energy. Withal, the companies which produce energy by using fossil resources like oil and gas are still powerful competitors in the energy sector. That is why many authors relate to the Positioning School. Conformable to KREWITT ET AL (2005) especially the long-term strategies are depending on the policies of the local governments. Because of that, the PV technology needs to improve its standing towards other technologies to keep financial supports.

With in the PV sector there are many possibilities for product differentiation which is, according to PORTER (1980), an important barrier to market entry. STEFFENS (2007) says that there are many different ways of integrating PV in buildings and that there are various ways to combine PV panels to a complete energy producing system. Companies can differentiate by offering more or less effective and efficient systems. Therefore, they need a lot of different suppliers to implement the PV panels into a PV System. A good position in the market depends on cooperation with the various producers of for example inverters, sub-constructions or cables.

2.6 Forecasting the Future

The DEPARTMENT TRADE AND INDUSTRY (DTI) (2001) is reporting about the future of the construction sector. They give their ideas of what the future will look like and what kind of changes could or should happen. Most authors agree that PV is the technology of the future because the sun is a never ending resource and global warming is on everybody's lips these days. FELLS (2002) thinks that the trend will go to renewable energies like PV because *"[t]he European Union plans to double its renewable energy portfolio from 6 per cent to 12 per cent by 2010"*. These expectations could motivate strategists within the PV sector to define strategies which include growth in terms of production capacity, recruiting more staff and expansion towards new sectors. RAGWITZ and HUBER (2006) are saying that *"[a]t EU level, the future importance of renewables has been recognised in different policy documents and energy strategies for more than ten years"*. They forecast that it will not be easy for companies in this sector to set long term targets which will be effective and do not implicate additional cost for the EU at the same time. This is because the development and future changing of the global and local energy markets are not conceivable. Furthermore, Problems of integrating electricity from renewable energies into liberalised energy markets are discussed by RAGWITZ and HUBER (2006). They argue that some technical constrains have to be addressed like *"[...] the quality of forecasting for fluctuating renewables, the responsibility for providing additional balancing power and the rules of grid extension and integration".*

2.7 Formulating the Strategies

This is a point in the strategic process where decisions must be made. Those responsible have to choose between alternatives to push on the process of changing. According to FEURER and CHAHARBAGHI (1997), *"[...] no simple strategy process or single strategic capability will lead to a sustainable competitive advantage"*. To meet the main aims of the company and their stakeholders, strategists should concentrate on the most important decisions. HARRISON and PELLETIER (1998) think that strategic decisions contain a lot of dynamic variables which make the formulation very complex. Therefore, *"[o]rgani[s]ations [have to] adjust dynamically their characteristics to the requirements of the environment by constantly changing their strategies and strategic capabilities"*. (FEURER and CHAHARBAGHI, 1997)

In the PV sector the main strategies focus on cost reduction and improving the PV-systems and the effectiveness of their components. WALZ (2006) argues that companies should increase their position in the global PV market, because the leadership as suppliers for this kind of technology is held by firms from Germany and Japan. Other companies from China, Brazil or India will enter the market in the near future. HELD ET AL (2006) conclude that the strategies should be aligning with growth, improvement of technical performance, enhancing social acceptance and reduction of cost.

2.8 Implementing the Strategies

"[...] [S]trategy formulation can no longer be separated from strategy implementation because of the speed which is necessary to exploit opportunities in the competitive environment". (FEURER and CHAHARBAGHI, 1997) The effectiveness of choice is dependent on the acceptance of the strategic choice by the relevant stakeholders in the external environment and within the organisation itself. In accordance with HARRISON and PELLETIER (1998), it is not enough to choose the best option without thinking about implementation. Strategist should provide plans to implement the strategy considering the internal structure of the company and the relationship to their environment. Everyone in a PV producing company needs to understand why and how the technology needs to be improved and what kind of changes will lead to success.

3 Conclusion

The purpose, aims and objectives of companies in the PC sector are comprehensible and the dependency on the internal and external environment is well reported in the literature. The mentioned schools of thought are practicable, although they are very complex and not easy to convert because of various influencing factors and forces. Some statements of the authors could be seen critically.

Porter (1980) argues that it is more important to focus on value rather on cost when companies try to analyse their competitive position. I would more agree with HELD ET AL (2006), because the cost is still a mayor problem of the PV technology, so companies in this sector need to focus on both. The EU COMMISSION (1997) thinks the same by saying that *"[g]rid-connected photovoltaic generation is not competitive at currently prevailing classical fuel-based generation costs [...]"*. Added value could accrue by additional *"[...] advantages of building integration such as lighting, heat supply [or] façade substitution [...]"*. (EU COMMISSION, 1997)

Another issue is that some authors see a main problem in the social background of the acceptance of the technology itself. People still face the topic PV sceptically. They do not accept new technologies for which additional prime cost are needed and the backflow of the money will be in the past. It seems to be very difficult to forecast the acceptance of PV in the population in the future. Therefore, financial incentives on the part of the government are necessary. I conform to the authors which say that this addiction is fundamental for the future of the PV technology. The arguments of the mentioned authors about forecasting the future of PV are plausible because companies just can make vague assumptions of what the development of the technology and the markets will be. But this does not mean that they should not be considered.

I am also sceptical about the competitive analysis from MCNAMEE ET AL (1999) because it is not easy at all to forecast the behaviour or strategies of your competitors. It is like the prisoner's dilemma which is reported by CLEGG ET AL (2005). There is always someone working against you and everyone 'keeps his cards close to his chest'. Even if the strategic benchmarking model would be realised you can not take everything for granted which is published by the companies in public.

Finally, the formulation and implementation of strategies are the very complex topics and critically discussed in the literature. I would agree to CHOO (2005) who argues that *"the idealistic approach seems to place strong emphasis on the ability to manage the complexity of external and internal environment in relation to the praxis of strategy".* The people who create the strategy should be involved in the implementation to guaranty that everyone understands and accepts the choice. In the PV sector strategies are often set out by associations or politicians who are not directly linked to the companies. This could cause problems because of the lack of relevant information in terms of needs and resources of the companies and their environments.

References

CHOO, K L (2005), *Examining the practice of strategic management in contemporary organisations.* In: *International Review of Business Research Papers*, Vol. 1, No. 2, Nov. 2005, pp. 100-112.

CLEGG, S, KORNBERGER, M, PITSIS, T (2005), *Managing and Organisations: An Introduction to theory and practice.* London: Sage.

COLE, G A (1997), *Strategic Management*, 2nd ed. London: Thomson.

DEPARTMENT TRADE AND INDUSTRY (DTI) (2001), *Constructing the Future – Foresight Construction Associate Programme (June 2001).*

EU COMMISSION (1997), *Energy for the Future: Renewable Sources of Energy – White Paper of a Community Strategy and Action Plan*, Communication from the Commission (26/11/1997).

FELLS, I (2002), *Clean and secure energy for the twenty-first century.* In: *Proc Instn Mech Engrs*, Vol. 16, Part A: J Power and Energy, pp. 291-294.

FEURER, R, CHAHARBAGHI, K (1997), *Strategic development: past, present and future.* In: *Training for Quality*, Vol. 5, No. 2, 1997, pp. 58-70.

Forschungsverbund Sonnenenergie (FVS) (2003), *Forschung eröffnet neue Horizonte für Solarzellen* [online], press report (25th Sept. 2003), URL:http://www.fv-sonnenenergie.de/publikationen/pm_03_09_25_jahrestagung_01.pdf (06/03/07).

HARRISON, E F, PELLETIER, M A (1998), *Foundations of strategic decision effectiveness.* In: *Management Decision*, Vol. 36, No. 2, 1998, pp. 147-159.

HELD, A, RAGWITZ, M, HAAS, R (2006), *On the success of policy strategies for the promotion of electricity from renewable energy sources in the EU.* In: *Energy & Environment*, Vol. 17, No. 6, 2006, pp. 849-868.

KELLETT, J (2003), *Renewable Energy in the UK Planning System*. In: *Planning, Practice & Research*, (Nov. 2003), Vol. 18, No. 4, pp. 307 – 315.

KEWITT, W, NITSCH, J, REINHARDT, G (2005), *Renewable energies: between climate protection and nature conservation*. In: *Int. J. Global Energy Issues*, (2005), Vol. 23, No. 1, pp. 29 – 43.

MCNAMEE, P, GREENAN, K, MCFERRAN, B (1999), *The competitive analysis model – a new approach to strategic development for small businesses*. In: *Benchmarking: An International Journal*, Vol. 6, No. 2, pp. 125 – 146.

PATEL, M R (1999), *Wind and Solar Power Systems*. London: CRC Press.

RAGWITZ, M, HUBER, C (2006), *Editorial-Status Quo and the future of renewable energies in Europe*. In: *Energy & Environment*, Vol. 16, No. 6, pp. v-viii.

LAURENS, S C (2004), Strategic Management Overview [online], URL:http://blog.laurenstravels.com/strategy/index.html (18/03/07).

SHEFFIELD, J (1999), *World population and energy demand growth: the potential role of fusion energy in an efficient world*. In: *Philosophical Transaction: Mathematical, Physical and Engineering Sciences*, The Approach to Ignited Plasma. (Mar. 15, 1999) Vol. 357, No. 1752, pp. 377 – 395.

STEFFENS, A (2007), *Photovoltaik – Energie der Zukunft*. Brochure (27/2/07), Weiler: Steffens Consulting.

VAN OVERSTRAETEN, R, MERTENS, R, LUTHER, J, LUQUE, A, BLOSS, W (1996), *Photovoltaics*. In: EUREC Agency, *The Future of Renewable Energies – Prospects and Directions*. London: James & James.

WALZ, R (2006), *Impact of strategies to increase RES in Europe on employment and competitiveness*. In: *Energy & Environment*, Vol. 17, No. 6, 2006, pp. 951-975.

WELFORD, R, GOULDSON, A (1993), *Environmental Management & Business Strategy*. London: Pitman Publishing.

WEISS, I, ORTHEN, S, STIERSTORFER, J, GISLER, R (2006), *Country Analysis – assessment of 12 national PV policy frameworks*. In: *PV Policy Group European Best Practice Report* (May 2006), pp. 17-59.